Early
NATURAL DISASTERS
Encyclopedias

DROUGHTS

by Laura Perdew

Early Encyclopedias

An Imprint of Abdo Reference
abdobooks.com

abdobooks.com

Published by Abdo Reference, a division of ABDO, PO Box 398166, Minneapolis, Minnesota 55439.

Printed in China.
102024
012025

Editor: Marie Pearson
Series Designers: Candice Keimig, Joshua Olson
Production Designer: Joshua Olson

Library of Congress Control Number: 2024938384

Publisher's Cataloging-in-Publication Data

Names: Perdew, Laura, author.
Title: Droughts / by Laura Perdew
Description: Minneapolis, Minnesota: Abdo Reference, 2025 | Series: Early natural disasters encyclopedias | Includes online resources and index.
Identifiers: ISBN 9781098296032 (lib. bdg.) | ISBN 9798384917038 (eBook)
Subjects: LCSH: Droughts--Juvenile literature. | Natural disasters--Juvenile literature. | Weather--Juvenile literature. | Environmental science--Juvenile literature. | Earth science--Juvenile literature. | Encyclopedias and dictionaries--Juvenile literature.
Classification: DDC 363.3493--dc23

CONTENTS

Defining Drought

A drought is a time of very dry weather. The area gets less rain and snow than normal. There is a water shortage. Lakes and rivers might dry up. Droughts cause harm to people, animals, and ecosystems. A drought is a natural disaster.

Where Droughts Happen

Droughts can happen anywhere. They look different in different places. A week without

Severe droughts can leave the ground cracked.

Boats can get stranded on banks when water levels drop during a drought.

rain is unusual in a rainforest. That might be a drought. But in a desert, a week without rain is normal. It would need more time without rain for a drought to happen.

Climate vs. Weather

Weather is what it is like outside at one time. That includes temperature, wind, moisture, and clouds. Climate is the average weather in an area over a long period of time.

Beginnings and Ends

Droughts begin slowly. They are not always easy to notice at first. That makes it hard to know when a drought begins. It is also difficult to know when a drought has ended.

FUN FACT!

The Atacama Desert is in Chile. It is one of the driest places on Earth. Some parts receive less than 0.2 inches (5 mm) of rain a year.

A lack of rainfall over an unusual length of time can be a sign of drought.

A Natural Disaster

Droughts do not look like other natural disasters. They happen gradually. Yet they have big effects. Living things need water to survive. Water shortages cause harm to people and the natural world.

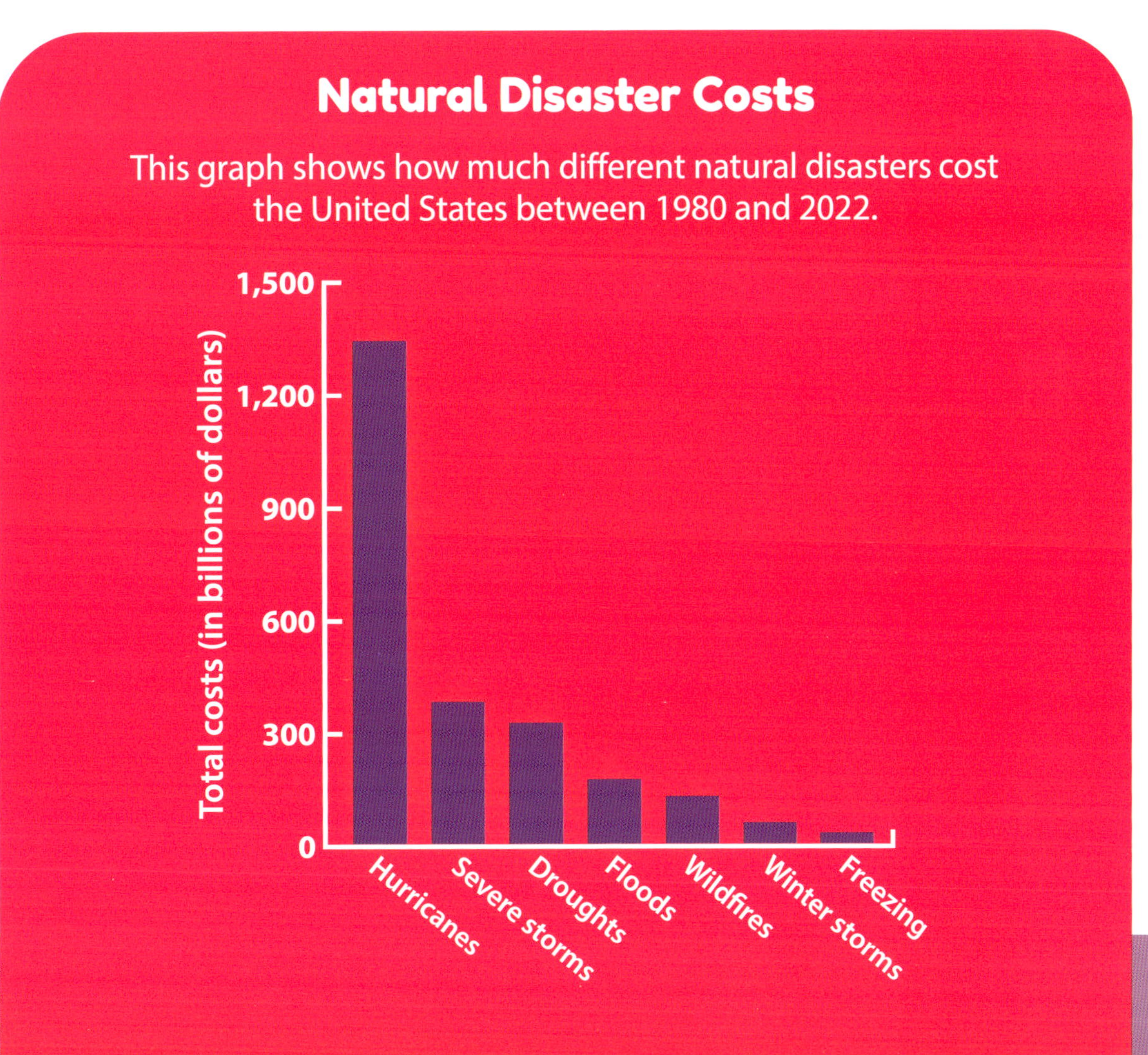

Satellite images show Earth's vast oceans.

Water on the Move

More than 70 percent of Earth is covered in water. Most of the water is in the ocean. Water is also found in lakes and rivers. It is frozen in glaciers. It falls as rain, hail, or snow.

Hidden Water

Much of the water on Earth is hidden. It is in soil. It is in groundwater. There is water in the air. Water is always moving. It is part of the water cycle.

The Water Cycle

The sun makes water evaporate. The water turns into a gas. It rises into the air. There, it cools and condenses. Clouds form. The water falls to Earth as rain or snow.

Groundwater can seep out of cliffs and form waterfalls.

When clouds have a lot of water, it can fall as rain.

Weather on the Move

Weather on Earth is always changing. Clouds carry water. Bands of strong wind move the clouds. The bands of wind are called jet streams.

Jet Streams

Jet streams move clouds around the globe. They carry warm and cool air. They bring precipitation. Precipitation is water that falls from the sky, such as rain, snow, or hail.

Patterns

The jet streams follow a pattern. The pattern shifts when seasons change. The shift moves the air currents over different routes. That changes the weather in an area.

Jet Stream Map

There are two types of jet streams. Polar jet streams are close to the poles. Subtropical jet streams are closer to the equator. Their exact paths are always shifting.

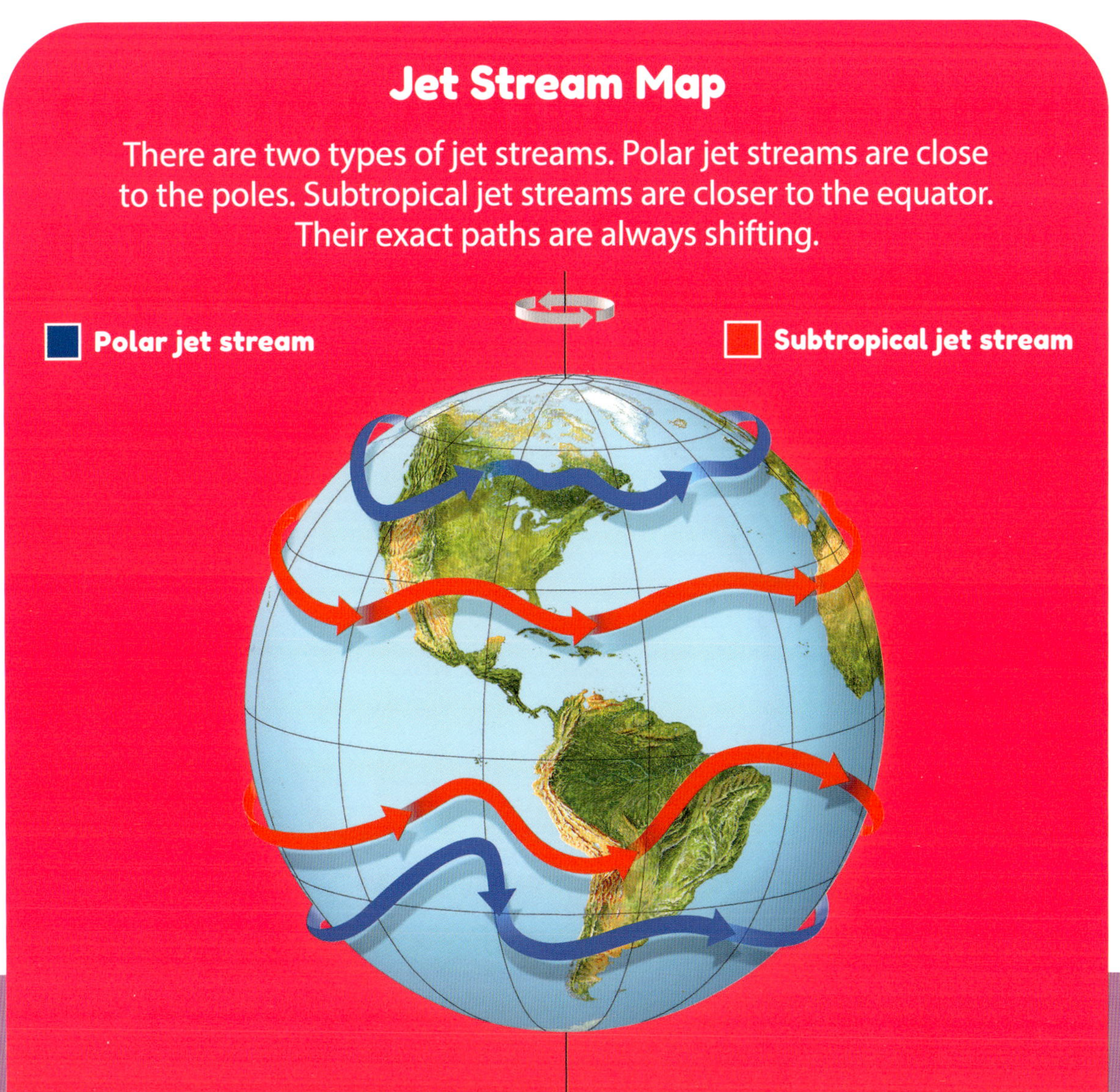

Natural Causes

Drought happens when there is less precipitation than normal. It gets worse if no rain or snow falls over a long time. Droughts can be a few weeks long. The longest are more than ten years.

Long droughts can kill trees.

Tanzania's normal dry season lasts from June to October.

Dry Seasons

The amount of rain or snow changes by season. Some places have a dry season. They usually get less rain at that time of year. But sometimes the dry season is drier than normal. Other times the dry season goes longer than normal. These are signs of drought.

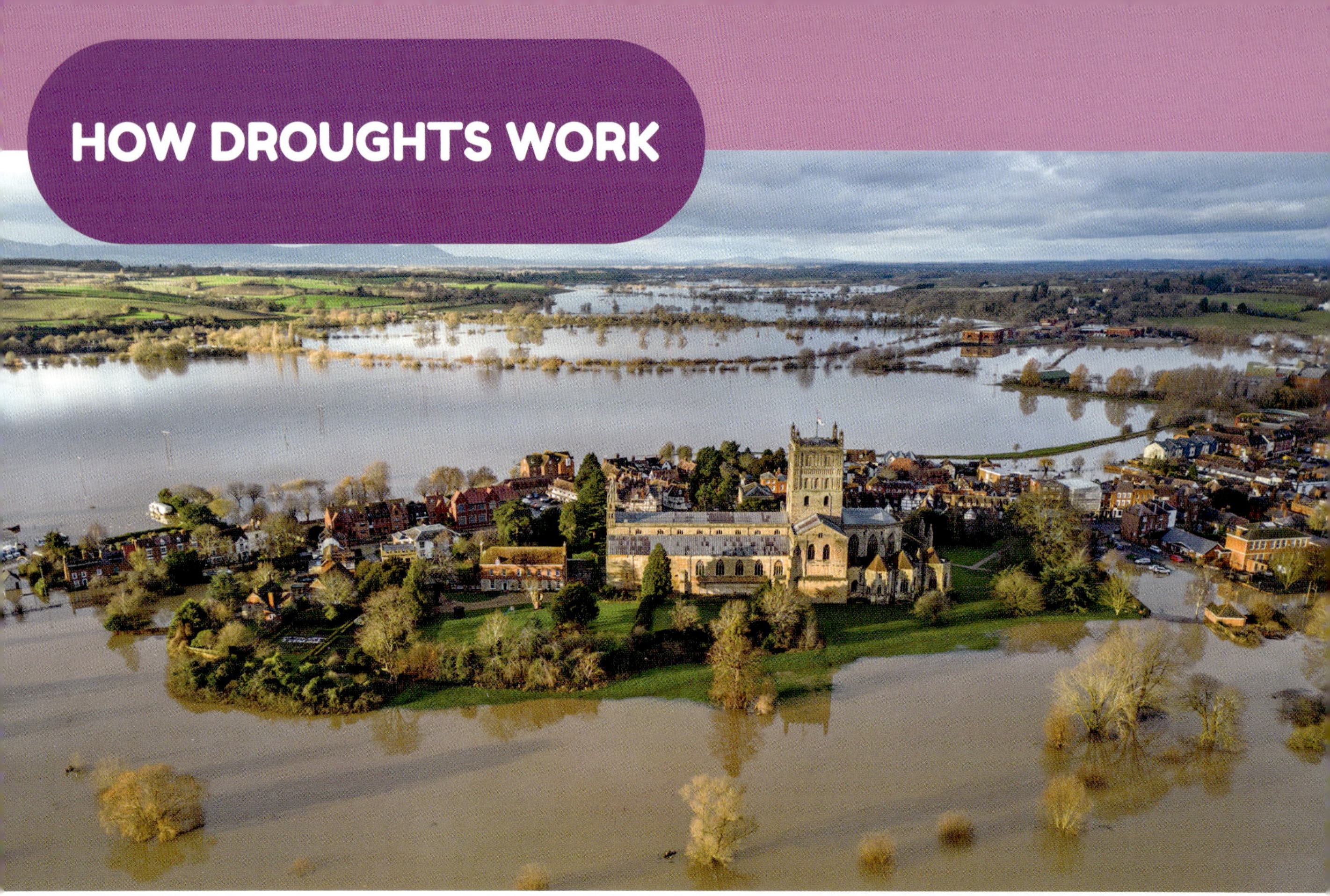

The jet stream caused major flooding in the United Kingdom during the winter of 2023–2024.

Changes in Weather Patterns

The jet streams move high above Earth. They go from west to east. Sometimes pockets of air block them. That makes clouds that bring rain go in different directions. These changes cause droughts in some places. They cause floods in others.

The Ocean

The ocean moves warm and cool water around the world. Changes in ocean temperatures affect weather and climate patterns. Longer and more frequent drought on land is linked to rising ocean temperatures.

Ocean Currents Map

The ocean is always in motion. The usual movements of its warm and cool waters are called currents.

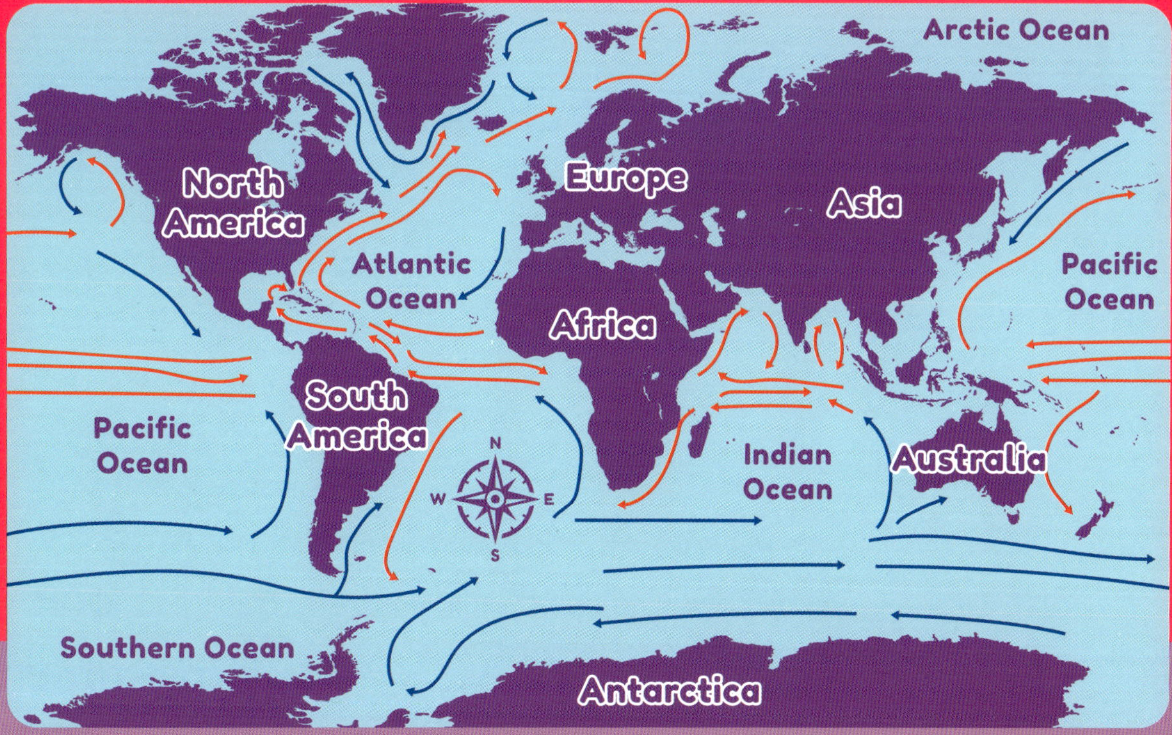

El Niño

El Niño is a natural weather event. It is when the water in the Pacific Ocean near the equator is warmer than normal. This causes more rainfall in parts of North and South America. It often causes drought in Australia and Indonesia. El Niño happens every two to seven years.

Roads in California washed away in August 2023 when storms driven by El Niño brought heavy rain.

La Niña

La Niña is the opposite of El Niño. It happens when the Pacific Ocean temperature is lower than normal. It causes changes in weather patterns too. Parts of North America may have droughts in a La Niña year.

Parts of the Amazon rainforest have been cleared for agriculture.

Deforestation

Human activity has made drought more common. Logging is one cause. Plants hold moisture. They release that moisture into the air. This helps cause rain. Tree roots help keep moisture in the soil. Without trees, less water stays in the area.

Agriculture

Healthy soil holds moisture. It has worms and microorganisms that move through the ground. Their movement creates holes that water seeps into. Healthy soil works like a sponge. But farming land without breaks from planting compacts the land. The soil cannot absorb water. It dries out quickly. Groundwater dries up.

If farmers do not take steps to keep their soil healthy, such as rotating crops, the ground will not soak up water well.

Demand for Water

The number of people on Earth is rising. Everyone needs water. People drink it. They use it to clean. Overusing water can cause drought. It can drain bodies of water.

Children aged 9 to 13 years old need five to six cups (1.2 to 1.4 L) of water each day.

Droughts can cause bodies of water that people rely on for water to dry up.

Climate Change

Human activity has caused rapid climate change. Climate change has shifted weather patterns. The changes cause more droughts in some places. The droughts last longer. They are more severe.

Some bodies of water have permanent water depth measurement sticks.

Identifying a Drought

Scientists look at many signs to know if there is a drought. They look at how much precipitation has fallen. Scientists check soil moisture levels. They measure the water levels, snow, and groundwater.

Other Signs

Plants wilt during droughts. Warmer temperatures over time can also be a sign. Scientists use all this data to know if a drought has started.

Snowpack

Snowpack is one important source of water. It is the amount of snow that collects during winter. It melts when warm weather returns.

Drought can cause plants such as ferns to change color before autumn arrives.

Predicting Drought

Scientists cannot know how long a drought will last. They cannot know how bad it will be. But tools let scientists see that a drought might happen.

Monitoring Drought

Many scientists work together to monitor drought. They gather data. They study it.

Scientists can usually predict a drought one month before it starts.

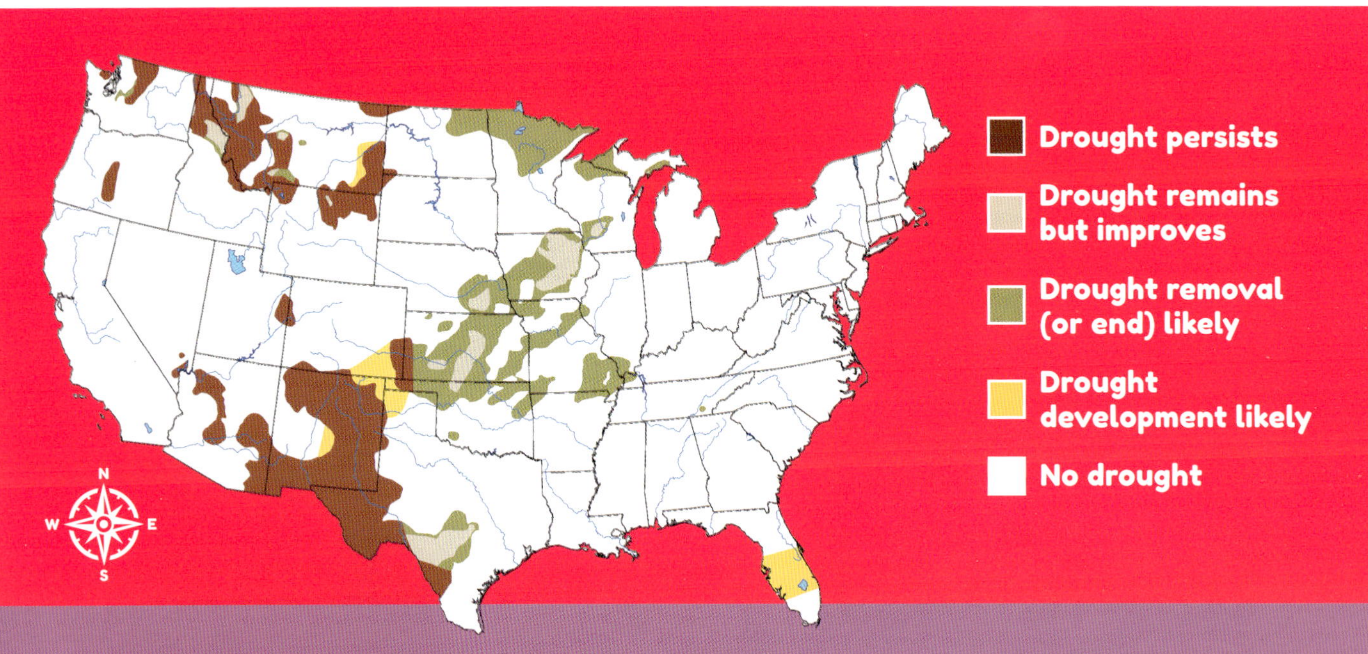

Researchers check trees for damage caused by not getting enough water.

Drought monitor maps show their findings. Early detection helps people take steps to reduce the damage of a drought.

Studying Trees

Tree rings show each year of growth. Scientists use the rings to study past droughts. Many rings close together can mean a time of drought.

Researchers use long poles to measure snowpack. It can be several feet deep.

DEWS

The Drought Early Warning Systems (DEWS) help predict drought. They use information collected by people from universities, tribal nations, and the US government. The data includes information about climate and weather trends. It also includes snowpack, streamflow, and soil moisture measurements.

Tracking Moisture

Some tools give a regular picture of the drought conditions in an area. The Vegetation Drought Response Index (VegDRI) posts a weekly US map. It shows moisture levels in plants.

FUN FACT!

Scientists sometimes ask regular people to collect data. This is called citizen science.

Researchers can use pressure chambers to measure plant moisture levels.

PDSI

The Palmer Drought Severity Index (PDSI) shows monthly drought conditions in the United States. It looks at precipitation, temperature, and evaporation. The PDSI data helps scientists predict the length and severity of a drought once

A PDSI map shows drought conditions in June 2021. Deep red is the driest, and dark green is the most moist.

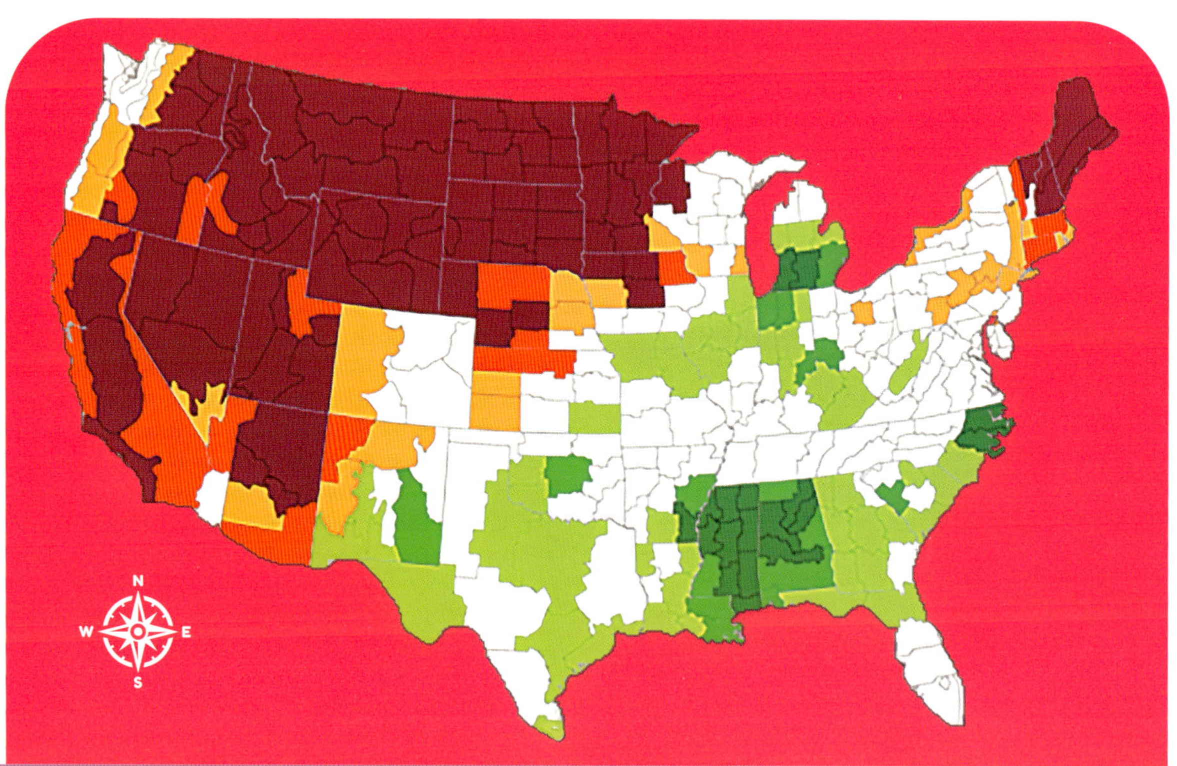

The SPI can compare droughts in regions that are very different from each other. One region may get more rain than the other but still have a drought.

it has started. It looks at rain but not snow. So it cannot give early warnings about drought.

SPI

The Standardized Precipitation Index (SPI) started in 1993. It is simpler than the PDSI. It measures precipitation only. The SPI can represent periods of drought from one month to three years long. It detects drought months earlier than the PDSI.

US Drought Monitor Map

The US Drought Monitor map shows five levels of drought.

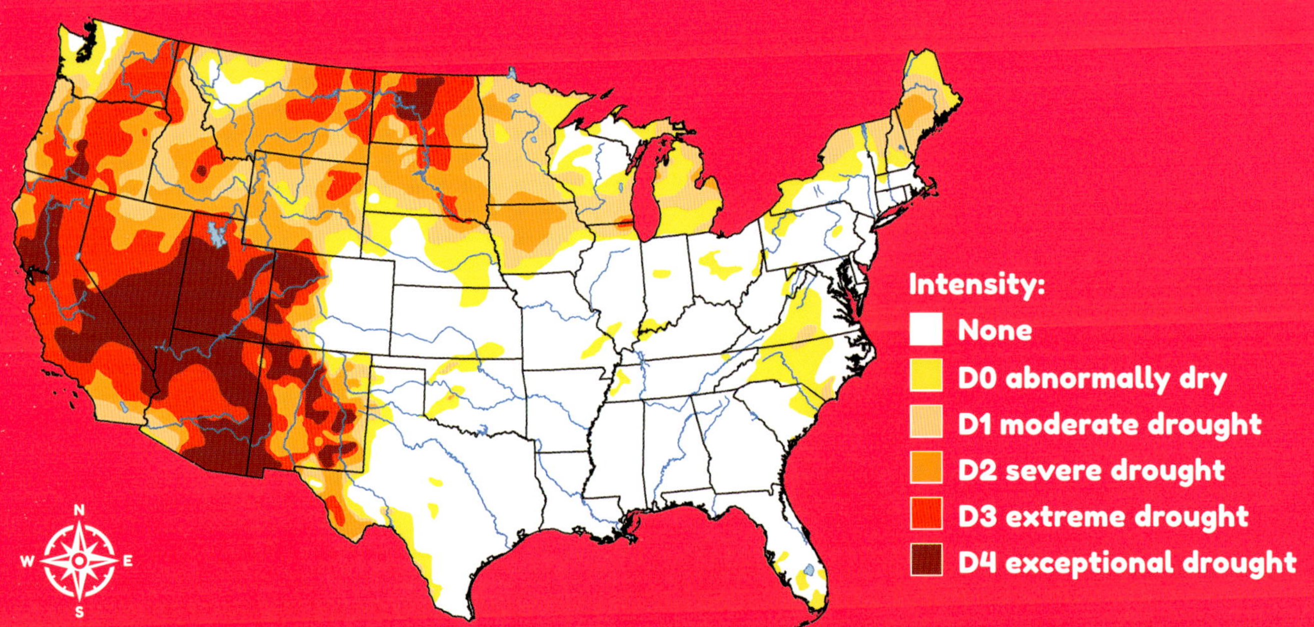

US Drought Monitor

The US Drought Monitor is a map. It shows US drought conditions. It started in 1999. Scientists make a weekly map. The map shows where drought is happening. Colors range from yellow to dark red. Dark red is the worst drought.

Categories

There are six categories on the map. The first is for normal conditions. Normal means there is no drought. The next is abnormally dry. The worst is exceptional drought.

A Utah Division of Forestry, Fire and State Lands officer used the US Drought Monitor map to show wildfire risk.

Using Satellites

Satellites help collect drought data. Some take pictures. Special cameras can show heat on Earth's surface. Warmer temperatures make water evaporate faster. This can cause drought. The images help scientists predict the amount of evaporation and the likelihood of drought.

NASA's Landsat program has several satellites that take photos of Earth. An artist illustrated the Landsat 8 satellite.

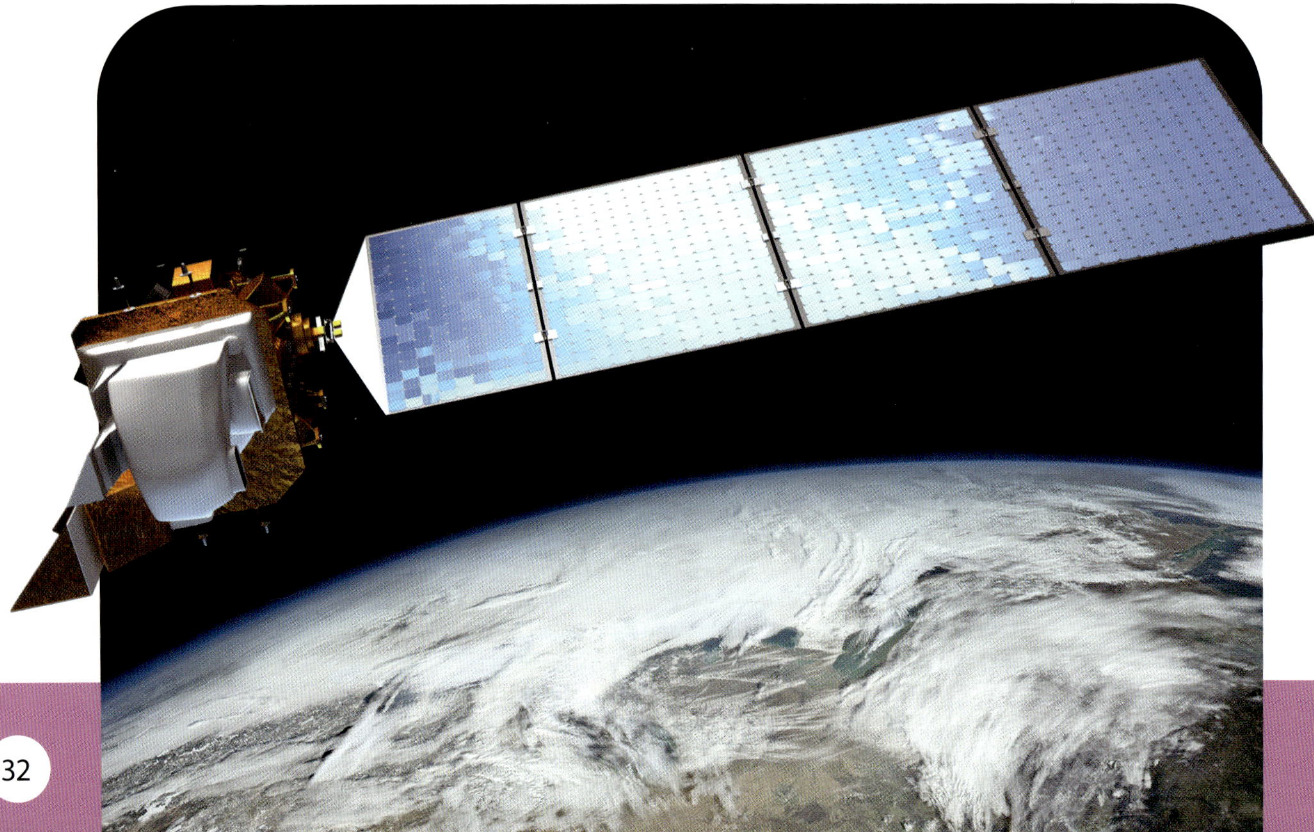

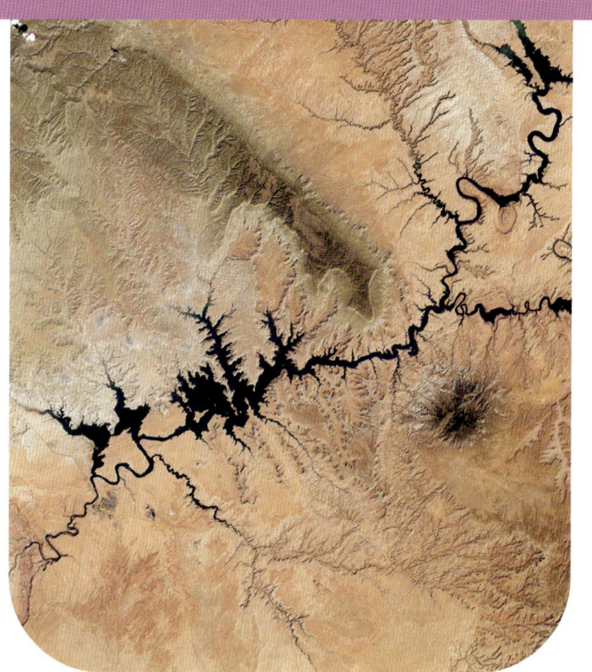

Satellite images show Lake Powell in Arizona in August 2017, *left*, and August 2022, *right*.

Other Satellite Images

Other satellite images track groundwater and streamflow. Some assess soil moisture or plant health. Satellites also track weather patterns. All this data helps scientists detect and monitor drought.

FUN FACT!

Some weather satellites circle Earth once every 24 hours. They are always looking at the same part of Earth.

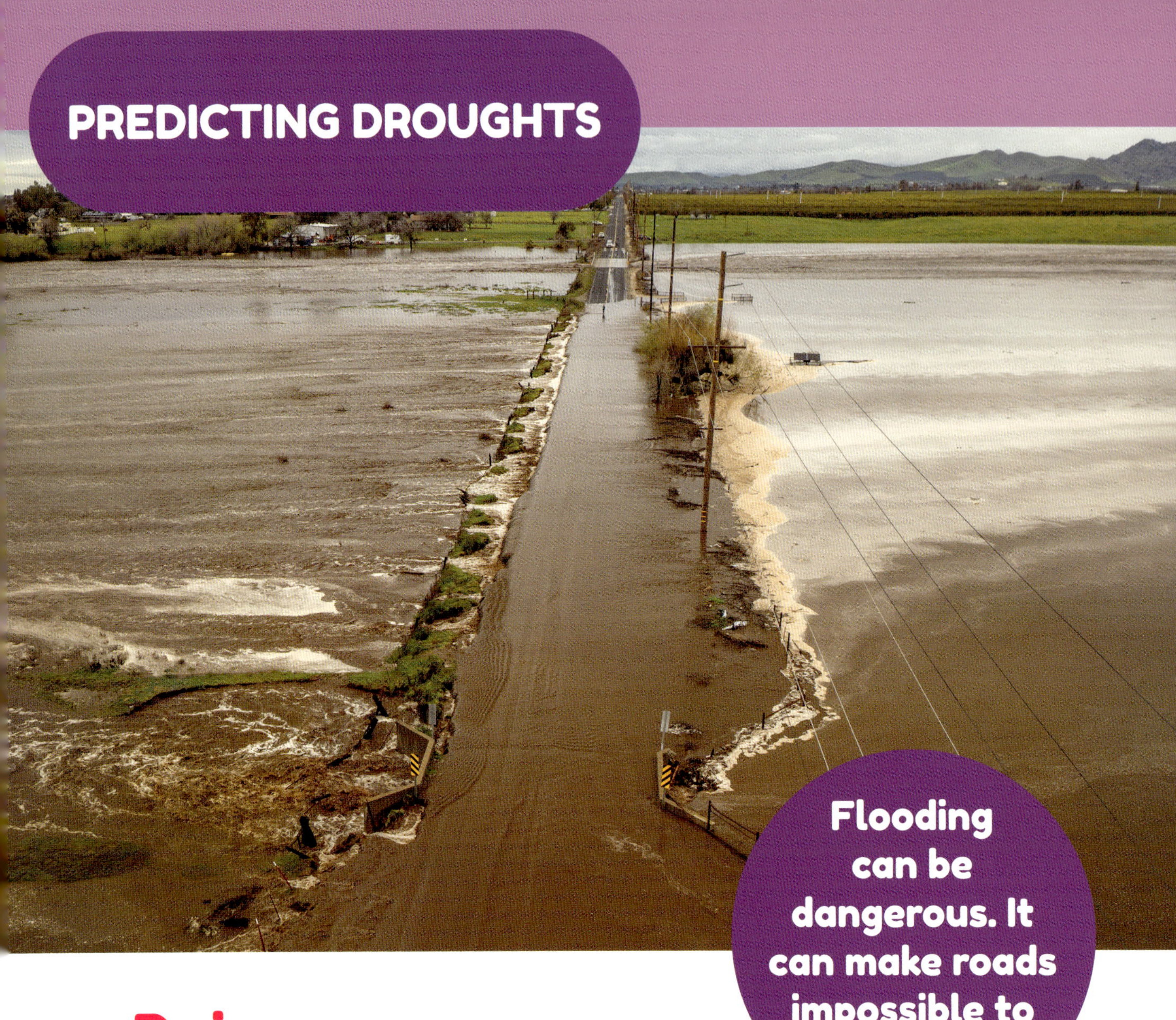

Flooding can be dangerous. It can make roads impossible to drive on.

Rain

A drought does not always end when it rains or snows. One rainfall provides only some relief. Too much rain may fall too fast. The dry ground cannot take in the water. It runs off into drains or streams.

The End of a Drought

Ending a drought takes time. It requires steady rain over days and weeks. Steady rain allows water to soak into the ground. This puts water back into the plants and soil. Groundwater and waterways refill.

A gentle rain will help plants after a drought.

Meteorological Drought

There are many types of droughts. Meteorological drought is less precipitation than usual. This kind of drought can start and end quickly. It can lead to hydrological drought.

Meteorological drought may be measured by an area getting less rain than usual by day, month, or season.

During a hydrological drought, the line of the usual water level for a lake or river may be visible.

Hydrological Drought

Hydrological drought is a lack of water in waterways. Water levels are lower in rivers and lakes. There is less groundwater and snowpack. This drought can lead to agricultural drought.

Flash Drought

Flash droughts begin or worsen quickly. They start after a period of less precipitation than usual. High wind, heat, and other weather changes make them grow worse quickly.

Crops may wilt if an agricultural drought happens.

Agricultural Drought

Farms need rainfall and groundwater. Agricultural drought is when there is not enough water for farms. It may keep farmers from planting. Drought stresses crops. It can kill them.

Socioeconomic Drought

The other types of droughts can lead to socioeconomic droughts. These droughts happen when the human demand for water is greater than the supply. People may not be able to get enough water, food, and other goods.

In 2022, drought dried up the tap water in the Monterrey, Mexico, area. People got water from trucks to do laundry and flush toilets.

Ecological Drought

Ecological drought affects nature. Wildlife numbers drop. Plants dry out. Ecosystems do not work properly. This leaves them vulnerable to disease and fire.

From 2018 to 2020, German forests experienced drought. Some trees died.

During a drought, water sources may dry up. Birds that eat fish may not have enough food.

Direct Effects

Nature is a balanced food web system. Each part of the system supports other parts. The system becomes unbalanced during a drought. It does not work properly. Everything in the system is affected.

Food Webs

Healthy food webs are systems of food chains. A food chain is the order of what eats what in an ecosystem.

Fish can get stranded on land if lakes dry up.

Harm to Nature

Plants cannot grow without water. They may dry out or die. That means some animals lose shelter from predators. They may not have enough food. Animals that live in the water lose habitat. Many animals are forced to move to other areas. Other wildlife does not survive.

Insect Invasion

Many insects do well during droughts. They swarm and eat plants. Drought-stressed plants are less likely to survive getting eaten. Some insects also move into buildings looking for water.

Locusts are a type of grasshopper. They sometimes gather in large groups called swarms.

In 2020, Ethiopia saw locust swarms caused by drought.

Plant Disease

Insects help spread diseases. Some plant diseases do well in hot, dry weather. Weak plants cannot fight off these diseases. Diseases can kill the plants.

Diseases can damage crops during droughts.

Dry plants burn quickly. Wildfires can easily start and spread.

Risk of Wildfire

Plants dry out during a drought.
Hot weather often comes with drought too. These factors make wildfires more likely. They also make wildfires larger.

Dust storms are also called sandstorms and haboobs.

Erosion and Dust Storms

Soil dries out during droughts. It blows away in the wind. Sometimes high winds stir up a wall of dust. The wall can reach high into the sky. This is a dust storm. It can turn the sky dark.

Air Quality

Wildfires and dust storms put ash, dust, and other tiny pieces of material into the air. These pieces get into the lungs. Breathing becomes difficult. Poor air quality can cause lung problems.

Dust storms limit how far a person can see. This makes driving dangerous.

Desertification

A dry area may be at risk of desertification during a severe drought. This means the area turns into a desert. Plants cannot grow.

After desertification, people need to care for the area so plants can grow again.

Flash flooding can move soil, changing the shape of the land.

Flash Flooding

Drought makes flash floods more likely. Dry soil is hard. The soil cannot take in water. When it does rain a lot, the rain causes flooding. This is called flash flooding.

In March 2024, the Sau Reservoir in Spain held just 1 percent of its capacity. Its dam provides energy to the region.

Energy

Some dams turn water movement into electricity. They cannot make energy if the water level is too low. Other types of power plants need water to cool them. Cool water keeps a power plant working safely. During a drought, it cannot make as much power. A power plant may need to shut down until the drought is over.

Water Supply for People

Drought lowers the water supply. People may have to cut back on how much water they use. They limit shower length. They use less water for cooking. Lawns may have to go without water.

Lawns may die during a drought if cities restrict water use.

Severe Water Shortage

Low water levels can make water quality worse. There may not be clean water to drink. There may also not be enough water for washing. This helps diseases spread. They make people sick. In extreme cases, people die.

When drought makes local water sources unsafe, nonprofit organizations may bring in clean water.

Finding Water

In some places, people must travel farther to get clean water during a drought. Women and children usually have that chore. They may spend time getting water instead of working or going to school.

THE EFFECTS OF DROUGHTS

Wheat is among the key global crops that have been under stress from droughts in the 2020s.

On Farms

As with other plants, drought makes crops more vulnerable to disease and pests. Even a week without rain affects farms. Farmers may have to plant fewer crops. Some crops may die.

Food Shortages

Drought can cause animal feed shortages. Livestock do not get enough to eat. Droughts cause human food shortages. People may go hungry. A lack of food affects health. People in poor health cannot fight off disease well.

FUN FACT!

A severe drought may have led to the collapse of the Maya civilization starting in 800 CE.

Rain after drought can cause tomatoes to crack.

Loss of Income

With less to sell, farmers lose income, which is money earned for work. There are also fewer jobs. Farm workers may be laid off. Workers in food plants may lose their jobs.

Farmers in Vietnam lost rice crops to drought in 2024.

People may no longer be able to afford certain foods when drought has caused a shortage.

Higher Food Prices

Having fewer crops limits the food supply. But the demand for food stays the same. So the cost of food goes up. People must spend more money to eat.

Supply and Demand

Supply is how much of something is available. Demand is how much of something people want. When the supply drops but the demand stays the same or goes up, the price rises.

Moving

Sometimes, droughts make places too dry for people to live there. People are forced to leave their homes. They move in search of clean water and food. They go in search of jobs.

Transportation

Low water levels can cause problems for boat travel. Some routes may close. Goods shipped

Although some people move away from drought, others move to places where droughts are common. Cheap housing is one of the reasons for this.

by boat must use another route. People must find other ways to travel.

The US Army Corps of Engineers uses a dredge boat named *Potter* to dig channels deeper in the Mississippi River when water levels are low.

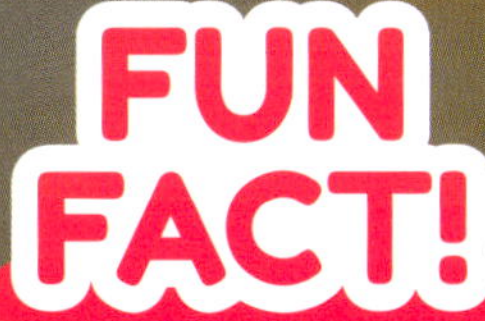

During a drought in 2013, engineers had to make a stretch of the Mississippi River deeper so big boats could pass through.

Stay Aware

Droughts might not seem very dangerous because they are hard to see. Tools such as the US Drought Monitor help people stay aware. These tools can warn of a drought before it happens.

City water towers sometimes have the city's name on them.

A worker monitors a Texas water treatment plant.

Where Does Water Come From?

Everyone should know where their water comes from. It helps people know how drought might affect them. Some water comes from reservoirs. Water may come from wells or other groundwater sources. Cities can store water in towers. Canals and pipelines supply water too.

When using a cup to rinse after brushing teeth, the water that is not used should not be dumped. It should be used for other purposes later.

Before a Drought

Saving water lessens the effects of drought. Everyone can conserve, or save, water at home. Every drop counts. Short showers use less water than baths. Turning off the faucet while brushing teeth saves water. Turning off taps tightly prevents dripping.

Other Home Tips

Water-saving devices on showerheads use less water. Low-flow toilets also conserve water. Compostable toilets do not use any water at all. Fixing leaky faucets and toilets saves water. Even leaks that drip slowly add up over time.

FUN FACT!

If a leaky faucet drips once a second, it wastes about 2,700 gallons (10,220 L) of water each year.

Some compostable toilets use wood shavings to help tiny living things break down the waste.

Outdoor Water

Lawns need a lot of water to stay healthy. Families can replace lawns with plants that need little or no water. Native plants are best. They are used to the natural water patterns. They do not need much extra water. Using native plants is most important in dry climates.

Instead of grass, people in dry climates that get droughts may plant cactuses and other desert plants.

A downspout sends water from the roof to rain barrels.

Rain Barrels

Homeowners can collect rainwater. Water from the roof runs into barrels. This water is then used for lawns and gardens. Some cities limit how much people collect. The limits ensure that some water returns to the local water cycle.

In springtime in Las Vegas, homeowners are allowed to water grass for 12 minutes a day up to three days per week.

Water-Smart Cities

Cities work to become drought-proof. They make water conservation plans. Some plans use low-flow toilets and faucets in homes and businesses. Some cities limit how much water people can use. Others make the whole water system work better with less waste.

Following a Water Plan

Some cities pay homeowners to replace lawns with native plants. Cities may pay residents or businesses to collect rainwater. Cities also use technology to detect pipe leaks. Fixing leaks quickly limits water loss.

Replacing lawns with native plants also helps native wildlife.

Gray Water

Using recycled water reduces the demand for more water. Toilets and landscaping can use gray water. That water is reused from sinks, showers, and more. Water used in manufacturing can be cleaned and reused too.

Gray water can be used to water plants.

California lawmakers approved turning wastewater into drinking water in December 2023.

Recycling to Drink

Some cities are recycling wastewater. They turn it into drinking water. In Oceanside, California, a new treatment plant cleans wastewater. The water goes through several cleaning steps. Then it is safe to drink. The process saves millions of gallons of water a day.

The plants on green roofs shade the building. This keeps it cooler.

Green Roofs

Urban areas are full of concrete, pavement, and other hard surfaces. When it rains, the water runs off into drains. It does not soak into the ground. Groundwater is not replaced. A green roof is a garden on a roof. It has soil and plants that collect water. This roof catches rainwater. It slows the water that does run off, lowering flood risks.

Green Streets

Natural areas on city streets also reduce runoff. They can be small gardens in a median. Parks, yards, and school lawns are other natural areas. They replace hard surfaces with natural ones. These natural areas slow down the water. The soil and plants take in the water.

The center of a roundabout can be a natural area.

Blue Roofs

Blue roofs collect and hold water. Blue roofs have basins to store water where it falls. Some have a pond. These roofs keep the water from becoming runoff. The water collected can be reused for irrigation, flushing toilets, and more. It cannot be used for drinking.

Sometimes green roofs also use blue roof basins, such as those filled with gravel.

Pavement that lets water seep through is called permeable pavement.

Water-Wise Pavement

Some cities use a pavement that allows water to seep through. They can use it for streets, sidewalks, and parking lots. Water soaks into the ground instead of running into drains.

Plants get up to 75 percent of the water from a sprinkler. But they get 90 percent of the water from drip irrigation.

Farmers

Farmers conserve water by using drip irrigation. The system delivers water to the base of the plant. Less water evaporates. Farmers monitor the soil. They see exactly when more water is needed. That way, they can deliver the right amount of water to the right plant at the right time.

Other Farming Methods

Some farmers collect and store water. Some build ponds. Others dig swales. These are low-lying stretches of land between ridges. Ponds and swales collect rainwater. Farmers use the water for crops. Farmers also plant crops that are resistant to drought. Growing crops adapted to the climate saves water too.

Irrigation ponds can store water for use during dry seasons.

Managing Soil

Another way farmers conserve water is by managing the soil. Healthy soil holds water and nutrients. It acts like a sponge. One way to keep soil healthy is to use compost. The compost puts nutrients back into the soil. Planting different crops in an area from year to year keeps soil healthy too.

Composting breaks down food scraps and plant matter. These materials turn into soil over time.

Crimson clover is a common cover crop.

Cover Crops

Farmers can plant cover crops after the main crop is harvested. Farmers also plant cover crops between the rows of main crops. This protects the ground from evaporation and erosion. Cover crops also help keep away weeds and pests.

Capturing rainwater for irrigation will help protect groundwater stores in China.

China

Some countries work to conserve water. China is trying to reduce the use of groundwater. Part of its plan is improving how water is used in farming. Some of the focus is on better irrigation techniques. China is promoting soil health too. Crop production has gone up. Groundwater use has gone down.

Israel

Israel fights drought by recycling water. More than 80 percent of its wastewater is recycled. In addition, Israel uses desalination to meet its water needs. The process turns salty ocean water into drinkable water. The country now gets about 80 percent of its water from desalination.

The Sorek desalination plant in Israel is one place that removes salt from water.

United Kingdom

The United Kingdom uses smart meters to save water. Residents track their water use online. The meters show how much water their households use. People can change their habits to better manage their use.

Smart meters are more accurate than older water meters. Smart meters also show usage immediately.

South Africa has removed invasive trees such as eucalyptus trees. These trees use much more water than native plants.

African Countries

Africa has frequent droughts. Yet many citizens are not aware of what they can do to conserve water. In schools in Cape Town, South Africa, students learn the importance of saving water. The Tanzanian government started a campaign to educate people. Education has helped both countries reduce their water use.

Replanting wetlands helps an area cope with flooding that may come after drought. Plants soak up the water.

Nature at Work

Natural areas can protect against drought. Balanced ecosystems provide water, food, and shelter for everything that lives there. Healthy ecosystems recover more quickly when natural disasters happen. They protect against drought.

The Great Green Wall

The Great Green Wall is an effort in Africa to restore the land. People are planting native trees. They are improving soil health. They are teaching farmers different methods. When the project is complete, the Great Green Wall will stretch across Africa. It will improve millions of lives.

People planted trees near Rosso, Mauritania, as part of the Great Green Wall.

Forests at Work

Forests absorb rain. They replace groundwater. They filter water, which improves water quality. Yet many forests have been cut down. With the trees gone, water is not stored. Protecting and replanting trees saves water.

Every year, people plant almost two billion trees. But they cut down about 15 billion trees a year.

Wetlands are an important habitat for many species of birds, fish, amphibians, and more.

Wetlands at Work

Healthy wetlands act like sponges. They slow and store water. Groundwater supplies refill. More plants and trees can grow. Green areas protect against wildfire. Restoring wetlands protects areas from drought.

Beavers

Beavers can block up a stream with logs. This floods the area, making a wetland.

Saving Water

When a drought is predicted, people should limit all water use. Shower water can be captured and used to water plants or flush toilets. Large buckets can be filled for washing.

In 2021, the Marin Municipal Water District in California handed out information and tools to help people save water.

Hot Days

Hot weather often comes with drought. People should stay out of the sun. It is important to stay cool. Drinking enough water is also important.

Satellite images can show heat waves. Yellow represents the hottest points on the map. This includes areas that were 140 degrees Fahrenheit (60°C).

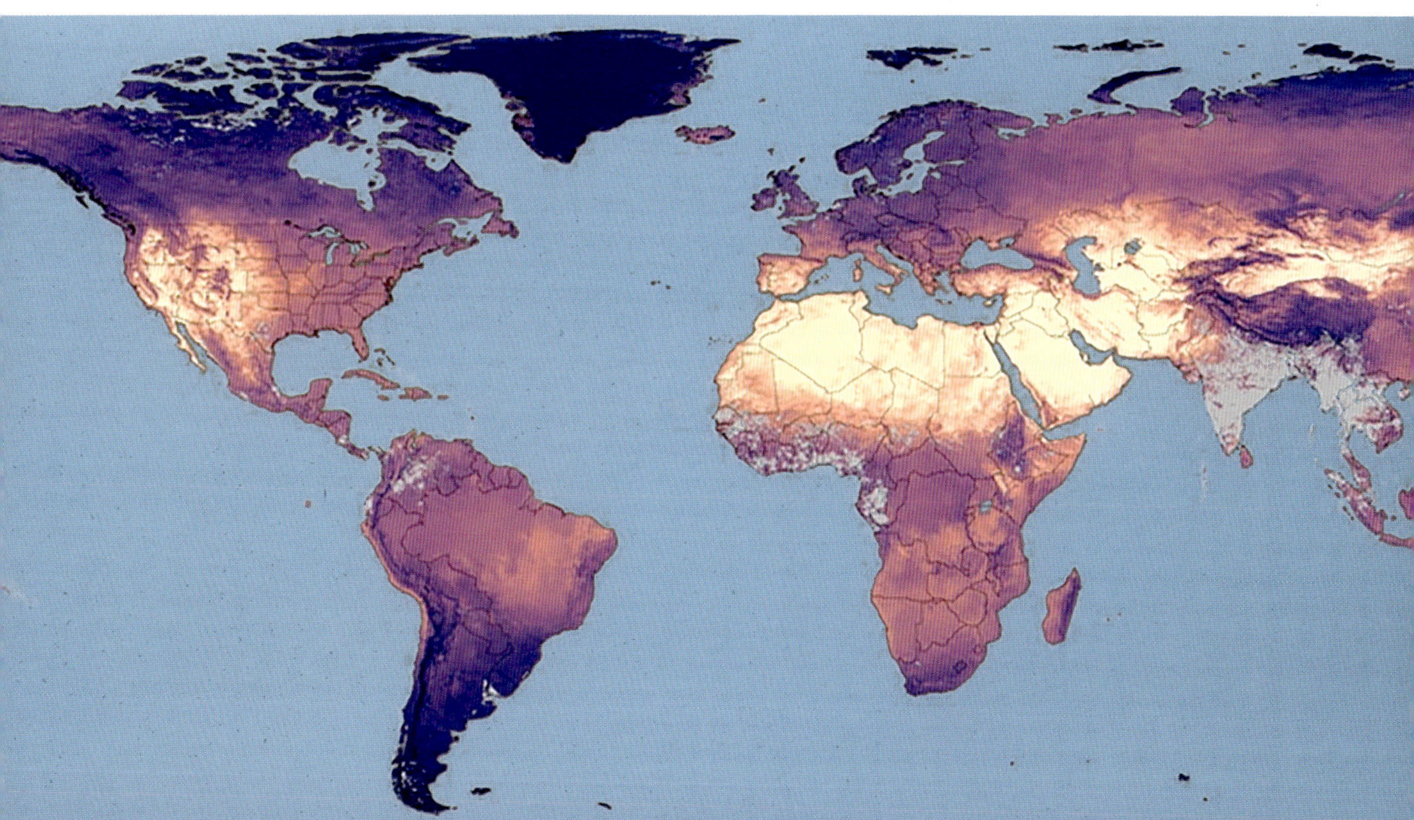

Wildfire Risk

Drought raises the risk of wildfire. Plants dry out. Sparks can cause a fire. Everyone must take care not to make sparks outdoors. That includes no campfires or cooking on a grill. It also includes not using matches outside.

People start about 85 percent of wildfires.

Evacuation Plan

People should know what to do if a wildfire starts. They should be ready to evacuate. They should plan evacuation routes. They should have emergency supplies on hand. If they evacuate, they should bring supplies for at least three days.

Evacuation routes can be slow because so many people are trying to leave at once.

Poor Air Quality

During a drought, there is more dust and dirt in the air than usual. Wildfires also add to poor air quality. Breathing poor air causes health problems. It affects the lungs and heart. People should stay indoors. Windows should stay closed. Home air filters should be replaced every 3 to 12 months.

Drought Plan

There are many steps people can take to keep themselves safe and save water during a drought.

Take short showers.

Wear mask to protect lungs.

Fix leaky pipes.

Water plants less, and water early in the morning.

Keep drought-resistant plants in yards.

Dust storms can affect cities in drought areas.

Masks

A face mask can be worn to keep dust, dirt, and ash out of the lungs. It should cover the nose and mouth. People with health problems may need an oxygen tank. The elderly, children, and people with health problems are most at risk. They should pay close attention to air quality alerts. The alerts provide information about how to stay safe.

Animals on the Move

Animals may move to cities during a drought. They are looking for food and water. People should stay aware of wildlife on the move.

About 29 percent of greenhouse gas emissions in the United States come from transportation.

Causing Changes

Changes in Earth's climate are normal. But these changes are now happening much faster than usual. The current climate change is caused by human activity. The main cause is burning fossil fuels.

Global Warming

Coal, oil, and gas are fossil fuels. They release carbon dioxide when burned. This gas traps heat in the air. Some amount of carbon dioxide is normal. It keeps the planet warm. But too much carbon dioxide has led to a rise in Earth's temperature.

Earth has warmed by about 2 degrees Fahrenheit (1.1°C) since 1900.

Climate Change Influence

Climate change has caused more frequent and more severe weather events. It has made droughts worse. Droughts are lasting longer. They are more extreme and hotter than in the past.

Droughts can cause formerly wide rivers to shrink to a trickle.

During a drought, more water evaporates or is used than is replaced by precipitation.

More Evaporation

Warmer temperatures make droughts worse. Evaporation happens faster. Plants and soil dry out. Water levels go down in rivers and lakes. Groundwater dries up.

Droughts have been happening more often in California since the late 1800s.

Shifting Storm Paths

Warmer temperatures move the jet streams. The shift changes the paths of clouds and storms. An area that usually gets rain gets a drought instead.

Water Cycle

Climate change disrupts the water cycle. More water evaporates. Some areas will lose water.

FUN FACT!

All water on Earth today is the same water that was on the planet billions of years ago.

But warm air also holds more moisture than cooler air holds. That can cause clouds full of moisture to pass over an area. Scientists predict that wet areas will get wetter because of climate change. Dry areas will get drier.

Many cattle ranches are in dry states in the Southwest. Droughts can mean there isn't enough grass for the livestock.

Less Snow

Snowpack is an important water source. Snow is frozen through the winter. It melts in warmer months. Melting snow sends water to rivers and lakes. But less snow falls in warm temperatures. With less snowpack, less water is available in drier months.

Satellite images showed the difference in snowpack in the Sierra Nevada in March 2010, *left*, and March 2015, *right*.

Melting mountain snow fills streams. These streams water lower areas.

Melting Sooner

Warmer temperatures also shorten winters. Snow melts earlier in the year. It melts faster. Waterways, soil, and plants are without water in the hotter, drier summer months.

The Austin, Texas, area issued a burn ban in August 2023 during a heat wave.

Drought in the United States

Since the early 1900s, droughts have become more frequent. In 2012, more than 80 percent of the United States saw drought. That was the worst since the 1950s. Temperatures are rising. Heat waves happen more often. Scientists expect that trend to continue.

Into the Future

Understanding droughts helps people prepare. This understanding makes people aware of the need for water conservation. It causes changes in farming. It inspires habitat preservation.

People who learn about droughts may be motivated to help restore natural areas.

What Was the Dust Bowl?

The Dust Bowl began in 1930. It lasted about ten years. The drought hit the Midwest and southern Great Plains regions of the United States. The area experienced high temperatures, a lack of rain, and strong winds. These combined with poor farming practices. Together, they made the Dust Bowl.

Settlers moved to the Great Plains because they could get land at no cost.

Wheat was a primary crop in the Great Plains heading into the Dust Bowl.

Settling the West

Before the 1930s, white settlers came from the eastern United States. They settled in the Great Plains. They started farms.

Farmers raised cattle on the Great Plains.

Dry Land

The Great Plains area was drier than the East. Much of the land was not right for farming. And many of the settlers had little farming experience.

Harm to Soil

Native grasses grew well in the dry conditions. But people replaced the native grasses with crops.

The land was plowed and planted over and over. The soil became drier.

FUN FACT!

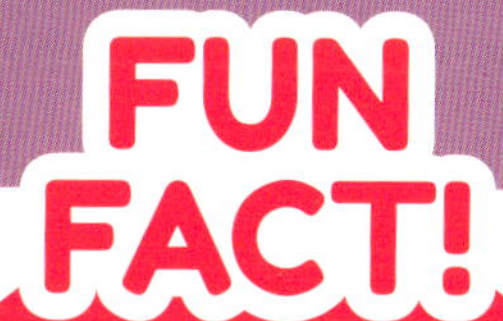

The Dust Bowl in the 1930s is sometimes called the Dirty Thirties.

With no grass to cover them, large areas of farmland were overexposed to wind and sunlight.

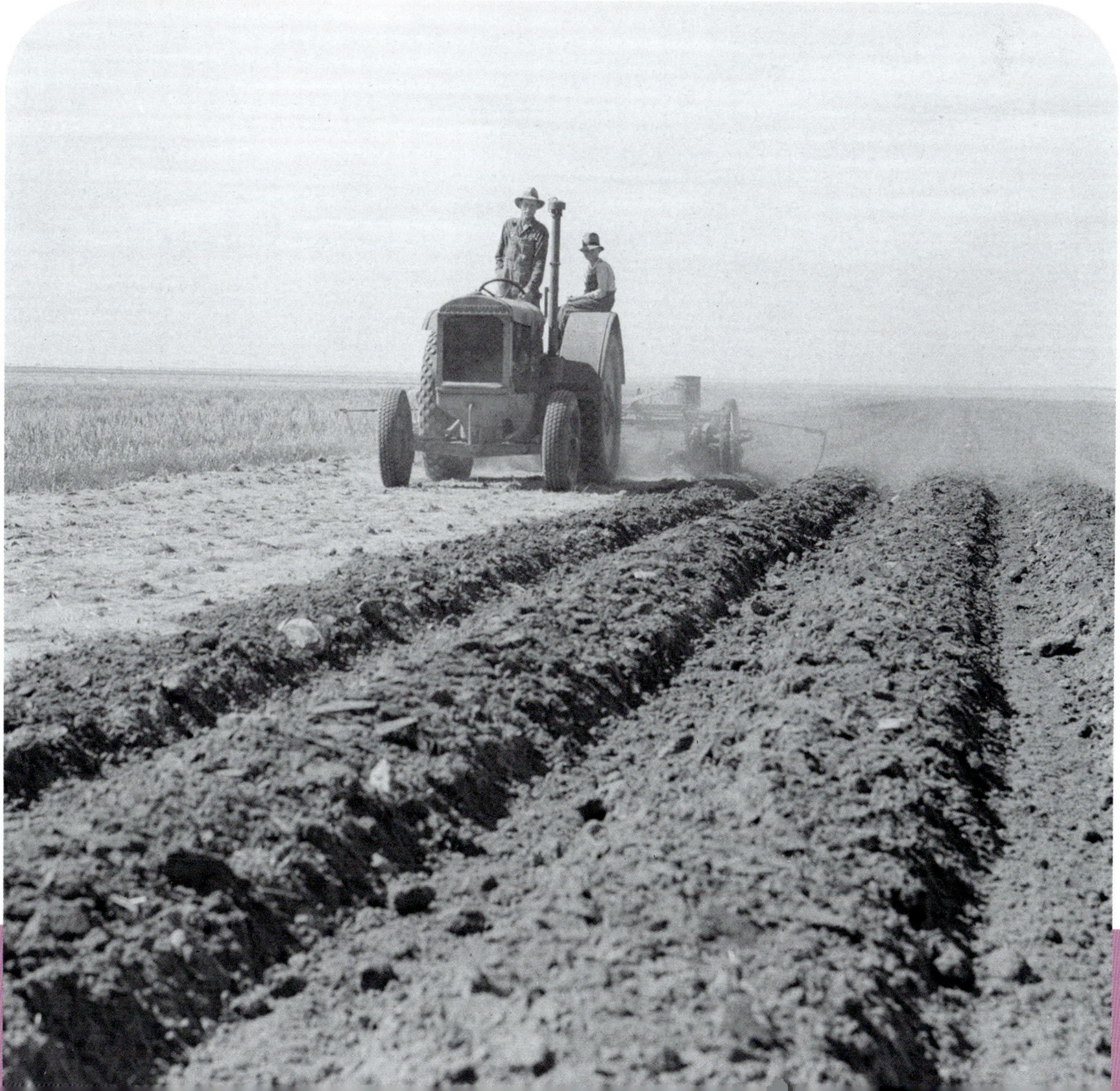

Dust Storms

Crops died when drought came in 1930. This left the ground uncovered. Dry topsoil began to blow away. Wind whipped across the plains. It stirred up dust. Huge dust clouds formed. Sometimes the dust was as thick as snow.

Some dust storms reached more than one mile (1.6 km) into the sky.

When dust got into the lungs, it could cause dust pneumonia. People with this condition had a fever, cough, and difficulty breathing.

Walls of Dust

The dust storms affected people's health. People and livestock died. The dust storms also hurt crops. Most crops failed. The stress of these conditions harmed mental health.

FUN FACT!

The dust storms were sometimes called black blizzards. That was because they looked like blizzards. But they had black material.

Dust covered machinery and drifted up against homes.

Migration

The Dust Bowl hurt the economy. Farmers lost income. More than two million farmers and their families moved. They needed to find work.

More Than One Event

Rain finally returned at the end of 1939. The drought had affected about 78,125 square miles (202,340 sq km) of land in the region. The economy did not recover until the 1950s.

Hit by Dust

Some of the dust storms were made of fine bits of soil. Others were sandy. Being hit with a sandstorm was like being rubbed with sandpaper.

A family from Oklahoma waited for a ride in 1936 so they could move to California.

Largest Rainforest

The Amazon rainforest is in South America. It is the largest rainforest in the world. Hundreds of waterways feed the Amazon River. Together, they make the world's largest drainage basin.

Amazon River Basin Map

The Amazon River Basin reaches into many countries.

Boats rested on dry Amazon riverbeds.

Wet and Dry Seasons

The Amazon normally has wet and dry seasons. But in 2023, a historic drought began in Brazil. Rainfall was the lowest in 40 years. Water levels dropped. They were the lowest in more than 120 years.

Effects on People

Many communities in the Amazon area use the rivers for transportation. But the rivers dried up. People were cut off from food and supplies. The drought killed crops. It led to a lack of clean drinking water. Many people became ill.

Only small boats could pass through parts of the rivers because of how low the water levels were.

Researchers studied the dolphins that died to find out what exactly caused their deaths.

Effects on Wildlife

The severe drought killed many fish. Dozens of freshwater dolphins died in one lake. The low water levels and high water temperatures caused these deaths. Dry conditions also put the forest at greater risk for wildfire.

River Dolphins

During the Amazon drought, Brazil's Lake Tefé reached 102 degrees Fahrenheit (39°C). High temperatures killed 153 freshwater dolphins in a single week.

AMAZON RAINFOREST DROUGHT OF 2023

Deforestation is responsible for about 10 percent of Earth's warming.

Causes

A loss of trees has made the Amazon rainforest hotter and drier. There are fewer trees to release moisture. So the air is drier. Dry air makes the remaining trees unhealthy. As a result, the rainforest is less able to recover from drought.

El Niño

El Niño in 2023 made the drought even worse. The weather event moved rain away from the area. Parts of the rainforest were unusually dry.

A man in Brazil dug a well in a dry riverbed to find water.

Climate Change

Climate change was another cause of the drought. Warming oceans pushed rain to other places. Warmer temperatures meant more evaporation. Plants, soil, and waterways dried up.

Manaus, Brazil, was covered in smoke on October 12, 2023, because of wildfires.

Firefighters worked to put out wildfires in Brazil.

Stuck in a Loop

The Amazon rainforest is stuck in a loop. Climate change makes natural dry seasons worse. Higher temperatures increase evaporation. This makes the forest drier. Drier forests are less healthy. They are more prone to wildfires. Fires lead to even more drying. The rainforest is less able to recover.

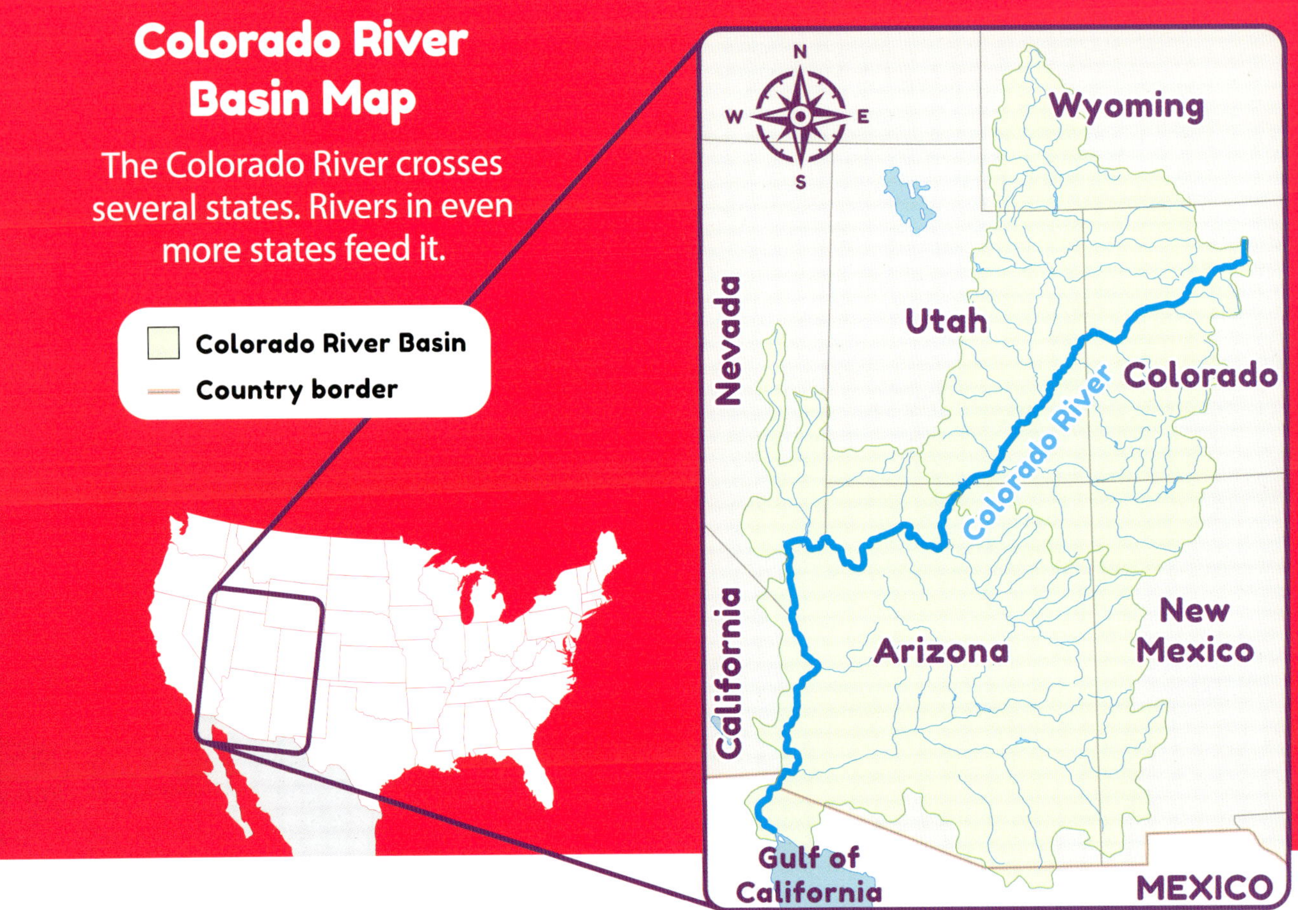

A Long River

The Colorado River begins in the Rocky Mountains of Colorado. It flows west and south for more than 1,400 miles (2,250 km). The river empties into the Gulf of California in Mexico. The river basin provides water to seven states.

FUN FACT!

There are 15 dams on the Colorado River.

Historic Lows

Since 2000, the region has suffered the worst drought in 1,200 years. Between 2000 and 2020, the Colorado River lost trillions of gallons of water. Water in reservoirs dropped by more than half. The river was barely a trickle when it reached the Gulf of California.

The flow of the Colorado River is down by 20 percent from its usual levels before 2000.

Causes

A hot drought is the cause of the Colorado River issues. A hot drought has below-average precipitation. It has higher than average temperatures. Because of climate change, the region's temperatures are hotter. Less snowpack forms. Evaporation is faster.

The Colorado River starts in Rocky Mountain National Park.

The Central Arizona Project canal runs for 336 miles (541 km). It brings water from the Colorado River to about six million people.

Demand for Water

The Colorado River has less water than normal. And there are more people in the region than ever before. More than 40 million people rely on the river for water. Farmers also depend on the river. About 70 percent of the river's water goes to agriculture.

Hundreds of miles of canals carry Colorado River water to nearby farms.

The Hoover Dam on the Colorado River provides electricity to about eight million people.

Effects

Less water is now available. Low water levels mean dams produce less electricity. States are making plans to cut back on water use.

Agriculture

Farmers in the Colorado River Basin get less water from the river. They cannot plant as many crops. Some fields are left unplanted. Farmers make less money. People in other farming jobs have lost work. Having fewer crops also means higher food prices.

Hopi Farming

Hopi people have farmed in the Colorado River Basin for more than 2,000 years. Their practices are based on tradition. They use little water.

In 2023, farmers of the Fort Yuma Quechan Tribe could get paid to not grow crops. The water they would have used stayed in the Colorado River.

Wildlife

The drought threatens habitats and wildlife in the Colorado River Basin. Fish numbers could drop with lower water levels. Wildlife on land also depends on the river. In addition, the drought stresses plants. They are more at risk of pest damage, disease, and wildfire.

Desert bighorn sheep are among the endangered species put at risk by drought.

Visitors enjoy rafting along the Colorado River's many scenic stretches.

Recreation

Many of the towns in the Colorado River Basin rely on tourism for income. People go boating, rafting, and fishing. But continued drought will mean less water for these activities. Fewer tourists will come. The areas will lose tourism income.

GLOSSARY

compost
A mixture of decomposed, formerly living matter that is used as fertilizer.

condense
To change from a gas to a liquid.

drainage basin
An area of land where all water flows downhill to a single point such as a river or its mouth.

economy
An area's system of goods and services.

ecosystem
A community of living and nonliving things and their environment.

equator
The imaginary line around the middle of Earth.

erosion
A process in which soil and rock are worn away and moved by wind or water.

evaporate
To change from a liquid such as water into vapor in the air.

fossil fuel
A source of energy that formed over millions of years, such as coal, oil, and natural gas.

irrigation
The methods farmers use to deliver water to crops.

reservoir
A place where water is stored for use.

runoff
Water that does not soak into the ground but runs into creeks, rivers, gutters, and sewers.

snowpack
Snow that collects during the winter, storing water until it melts.

water cycle
The ongoing movement of water on Earth.

TO LEARN MORE

More Books to Read

Buckey, A. W. *Weather*. Abdo, 2025.

London, Martha. *Droughts*. Abdo, 2020.

Weather. DK, 2022.

Online Resources

To learn more about droughts, please visit **abdobooklinks.com** or scan this QR code. These links are routinely monitored and updated to provide the most current information available.

INDEX

PHOTO CREDITS

Cover Photos: J. J. Gouin/Shutterstock Images, front (corn); Shutterstock Images, front (desert); Alexander Ermolaev/Shutterstock Images, front (girl); Natali Sam/Shutterstock Images, back

Interior Photos: Rachen Stocker/Shutterstock Images, 1; Shutterstock Images, 3, 5 (bottom), 9 (top), 13, 18, 21, 23 (top), 25 (bottom), 29, 37 (bottom), 41, 45, 47, 49, 57 (top), 57 (bottom), 62, 66, 80, 84, 85 (top), 90 (shower and mask), 90 (pipe), 90 (watering can), 90 (cactus), 92, 93, 101, 109 (top), 111, 113 (bottom), 114, 123 (top); Ami Vitale/Getty Images News/Getty Images, 4; Sebastian Castelier/Shutterstock Images, 5 (top); Delmotte Vivian/iStockphoto, 6; Red Line Editorial, 7, 15, 110, 118 (left), 118 (right); NASA, 8, 32; National Park Service, 9 (bottom); John D. Sirlin/Shutterstock Images, 10, 46; Gary Hincks/Science Source, 11; Sawat Banyenngam/Shutterstock Images, 12; Anna Barclay/Getty Images News/Getty Images, 14; David McNew/Getty Images News/Getty Images, 16, 34; George Rose/Getty Images News/Getty Images, 17, 125; Tony Savino/Shutterstock Images, 19; Ermolaev Alexander/Shutterstock Images, 20; Kyle Grillot/Bloomberg/Getty Images, 22; Patrick Pleul/dpa/picture alliance/Getty Images, 23 (bottom); Adam Hartman/NCEP Climate Prediction Center/NWS/NOAA, 24; Thomas Trutschel/photothek/picture-alliance/dpa/AP Images, 25 (top); Rich Pedroncelli/AP Images, 26, 27; NOAA, 28; US Drought Monitor, 30; Rick Bowmer/AP Images, 31; Joshua Stevens/NASA, 33 (left), 33 (right); iStockphoto, 35, 68; J. J. Gouin/Shutterstock Images, 36, 44; Marisa Estivill/Shutterstock Images, 37 (top); Bildagentur Zoonar GmbH/Shutterstock Images, 38; Gary Coronado/Los Angeles Times/Getty Images, 39; K. I. Photography/Shutterstock Images, 40; Zaid Bin Yaseen/Shutterstock Images, 40–41; Ververidis Vasilis/Shutterstock Images, 42; Minasse Wondimu Hailu/Anadolu Agency/Getty Images, 43; Feature China/Future Publishing/Getty Images, 48; Lluis Gene/AFP/Getty Images, 50; Suzanne Tucker/Shutterstock Images, 51; Mohammed Hamoud/Getty Images News/Getty Images, 52; Ashish Vaishnav/SOPA Images/LightRocket/Getty Images, 53; Jasper Suijten/Shutterstock Images, 54; Catherine L. Prod/Shutterstock Images, 55; Nhac Nguyen/AFP/Getty Images, 56; Andy Dean Photography/Shutterstock Images, 58; George Stringham/US Army Corps of Engineers/DVIDS, 59; Felix Mizioznikov/Shutterstock Images, 60; Brandon Bell/Getty Images News/Getty Images, 61; Stephen William Robinson/Shutterstock Images, 63; Aaron Finn/Shutterstock Images, 64; Anton Dios/Shutterstock Images, 65; Julia Rubin/AP Images, 67; Terry Chea/AP Images, 69; Lucas Eduardo Benetti/Shutterstock Images, 70; Bob Janssen/Shutterstock Images, 71; Cheng Wei/Shutterstock Images, 72; Michael P. Farrell/Albany Times Union/Hearst Newspapers/Getty Images, 73; Sirichai Chinprayoon/Shutterstock Images, 74; Justin Sullivan/Getty Images News/Getty Images, 75, 86; Kaca Skokanova/Shutterstock Images, 76; Jennifer Larsen Morrow/Shutterstock Images, 77; Ding Lei/Xinhua News Agency/Getty Images, 78; Gil Cohen Magen/Xinhua News Agency/Getty Images, 79; Peter Titmuss/UCG/Universal Images Group/Getty Images, 81; Chaideer Mahyuddin/AFP/Getty Images, 82; Shigeki Tao/Yomiuri Shimbun/AP Images, 83; Profi Trollka/Shutterstock Images, 85 (bottom), 91 (bottom); Jesse Allen/NASA/DVIDS, 87; Passakorn Sakulphan/Shutterstock Images, 88; David Odisho/Bloomberg/Getty Images, 89; Peter Parks/AFP/Getty Images, 91 (top); Solmaz Daryani/Shutterstock Images, 94; Ethan Miller/Getty Images News/Getty Images, 95; Mario Tama/Getty Images News/Getty Images, 96, 97, 123 (bottom); Jesse Allen/NASA, 98 (left), 98 (right); Maxim Petrichuk/Shutterstock Images, 99; Jordan Vonderhaar/Bloomberg/Getty Images, 100; Arthur Rothstein/Library of Congress, 102; Dorothea Lange/Library of Congress, 103, 105; Universal History Archive/Universal Images Group/Getty Images, 104, 107; Everett Collection/Shutterstock Images, 106, 108; Circa Images/GHI/Universal History Archive/Universal Images Group/Getty Images, 109 (bottom); Bruno Zanardo/Getty Images News/Getty Images, 112; Gustavo Basso/NurPhoto/Getty Images, 113 (top), 116; Michael Dantas/AFP/Getty Images, 115, 117; L. Powell/Shutterstock Images, 119; Gypsy From Nowhere Images/Alamy, 120; Manuela Durson/Shutterstock Images, 121; Sean Pavone/Shutterstock Images, 122; Maxwell McCammon/Shutterstock Images, 124